DINOSAUR ISLAND

AT SPACETIME PARK

NEW BRAUNFELS TEXAS

WWW.SPACETIMEPARK.COM

Dinosaur Island at SpaceTime Park

by Ron. L. Toms

© 2017, Ron L. Toms

Published by

RLT Industries,

New Braunfels, TX.

USA

All rights reserved. No part of this publication may be reproduced, stored or transmitted by any means without the express written permission of the publisher RLT Industries or the author Ron L. Toms.

ISBN 978-0-9776497-8-5

http://www.SpaceTimePark.com

Contents:

Stegosaurus	3
Hypsilophodon	5
Triceratops	7
Parasaurolophus	9
Insects	11
Amphibians	13
Pteranodon	15
Deinonychus	17
Reptiles	19
Minmi Ankylosaur	21
Pterodactyl and Velociraptor	23
T-Rex	25
Apatosaurus	27

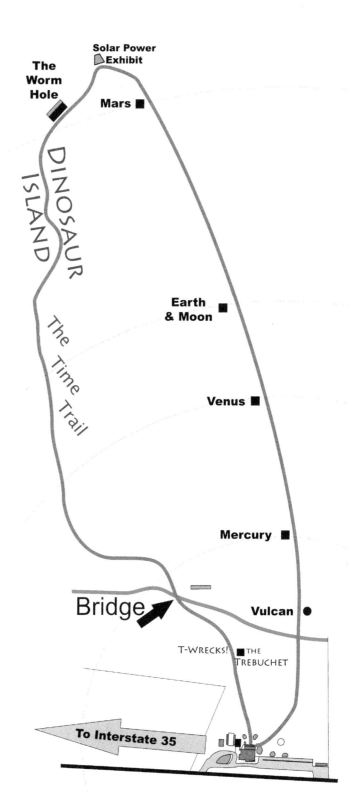

Is this a Stegosaurus?

You might think this is a stegosaurus, but you'd be wrong. It's actually a hesperosaurus, a slightly smaller relative of the stegosaurus. A full sized stegosaurus would be about twice this size. Over the hundred million years or so that these animals lived, there were a lot of variations in size for the stegosaurus-like family of animals.

Most of the stegosaurs lived in western north america in the late jurassic period, around 150 million years ago. But a few have been found in other places all over the world.

Stegosaurus

We know from the structure of their bones and how the muscles attach to the bones that these animals had a top speed of only about four miles an hour. You could easily outrun them, but you wouldn't have to. They're not predators. Look at the mouth. It has a beak like a bird. That means these animals probably ate a lot of seeds and nuts, and the occasional insect that wandered too close. But keep in mind, the insects of this era could be up to six feet long. Imagine a six foot centipede or a six foot scorpion being eaten by this animal. Now that would be a sight to see!

But the stegosaurus and hesperasaurus family of animals were also food for other predators. They had to defend themselves. Those tail spikes are called thagomizer spikes. They were used to defend against attacks by allosaurus dinosaurs.

No one really knows what the plates along the spine were for. They have large vein structures in them, suggesting a significant amount of blood flow — not something you want in a defensive structure that's likely to be damaged in battle. But the plates are not their biggest mystery. These animals appear to have had two brains! A tiny one in the skull, about the size of a large walnut, and a second one about twenty times larger at the other end of the spine at the base of the tail. But we're not really sure if that was a brain or some other kind of neurological organ.

in spite of all we know about them, the stegosaurus is still a mysterious animal.

Hypsilophodon

Look across the creek to see the hypsilophodon. When this animal was originally discovered we thought it was just a baby iguanodon, but later we realized it should be classified as its own species.

This animal lived about 130 million years ago, right before avian, or bird-like dinosaurs began to appear in the fossil record. Notice how the body style looks like a cross between a lizard and a bird. It even has a beak. Some scientists think this was a transition species caught in the process of evolving from reptile into a more avian form.

They were not very big. This is a full-sized specimen. But they were fast runners and had another fascinating feature. One thing that sets humans apart from most animals is our opposable thumbs for grasping things. This animal had opposable pinkies! We think they were used for grabbing and holding food. Maybe they were eating giant worms...

The Hypsilophodon fossils are only found on the isle of Wight off the coast of England and no where else in the world, which is exactly what you'd expect from a transition species. They change only where there is environmental pressure to change. But this is not a rare fossil, almost 100 specimens have been found so far.

Triceratops

Of course, you recognize this dinosaur as the Triceratops. These mighty creatures were among the last of the dinosaurs. They lived in the late cretaceous period, about a hundred million years after the stegosaurus, apatosaurus and allosaurus dinosaurs went extinct.

The Triceratops and Tyrannosaurs lived at the same time. Many paleontologists think they were fierce enemies. Fossils of each have been found with lethal wounds inflicted by the other. Since we didn't want anyone to get hurt, we had to separate these two on opposite sides of our park.

The big frill around the back of the triceratops skull was probably used to shield the vulnerable neck from attacks by smaller T-Rexes or other predatory dinosaurs. It may have also been an ornamental display for mating and species identification.

This one is just a baby. Full sized triceratops were almost three times as big as this one, up to 30 feet long (9 meters) and weighing five tons, much bigger than any elephants or mammoths. They were mainly plant eaters who had both a beak and teeth— a *lot* of teeth. They grew new teeth continuously throughout their lives (like modern sharks do) and typically had a hundred or more teeth in their mouth at any time.

Triceratops are one of the most common dinosaurs in North America. Over 47 new skulls were found between 2000 and 2010 in Wyoming alone.

Parasaurolophus

This is another North American dinosaur. These animals lived about 75 million years ago. They could walk on either two legs or on all fours, so this may have been another transition species caught evolving between quadruped and biped body styles.

That large crest on top of its head is its most interesting feature. It was made of the same kind of bone as noses are made from, and although air could pass through it, it was not used for breathing. The channels or tubes inside the crest ran from the nostrils all the way to the end and then came back to the nasal cavity just behind the nostrils. But their skulls also have the normal nasal cavities used for breathings.

Some crests were straight and long, and others were shorter and more curved, and They seem to have changed with age and gender. We're not sure what they were for, but most scientists think they may have been used like a trumpet to make distinctive sounds for attracting mates or to scare predators.

Insects

The Dinosaurs may be extinct, but not everything that lived in the time of dinosaurs is gone. Insects are older than dinosaurs by a few hundred million years, and they're still here. Of course, insects today are a lot smaller than they were back then. These insects in our display aren't even the largest that ever existed. Centipedes and scorpions grew up to six feet long and there were spiders the size of small dogs. Think about that the next time you go camping.

So why did insects get so large back then, and why don't they get that big today?

You may have heard of Global Warming. That's the idea that our atmosphere is changing and trapping more of the sun's heat. In fact, we can measure a lot more carbon dioxide in the atmosphere today than there was 100 years ago. But humans aren't the only thing affecting the atmosphere. Before the time of the dinosaurs the air was a lot different than it is today. 200 million years ago the air was a lot thicker with even more carbon dioxide and almost twice as much oxygen too. The denser, warmer and wetter atmosphere made the animals who lived then different too.

Insects don't breathe like other animals. Their blood doesn't transport oxygen. Instead, they absorb oxygen directly through pores in their skin. If they grow too big, the oxygen can't get to the deepest parts of their bodies. 200 million years ago a higher air pressure with more oxygen made it possible for insects to grow so large. If we brought one of these giant bugs into today's world it would suffocate within minutes in our relatively thin, dry, cold air. So you see, you have nothing to worry about!

Amphibians

Another group of animals older than dinosaurs is the amphibians.

Amphibians are born as water-breathers but change to breathe air as they mature. Because of this, biologists think amphibians were the first vertebrates to crawl out of the oceans to inhabit the land.

The ancient origins of frogs, salamanders and other amphibians begin to appear in the fossil record about 365 million years ago. Thats 150 million years before the dinosaurs appeared. Amphibians were the first real predators of the insects, and the progenitors of the family of reptiles which eventually became the dinosaurs.

Pteranodon

Look across the creek at the flying pteranodons. Although they lived at the same times as the dinosaurs, they are not part of the dinosaur family. These were flying reptiles.

They ranged in size in size from about as big as a crow up to the size of a small airplane, around 25 feet from wingtip to wingtip — larger than any flying creature since then. But the giants were rare. Most pteranodons were the size of these models or smaller.

Deinonychus

Don't be fooled, this is not a velociraptor! It's a deinonychus. But it's ok if you were confused. We've been told that when Michael Chriton wrote his book *Jurassic Park* he loved the name Velociraptor so much he wanted it to be a character in the story, but unfortunately, the velociraptor was way too small to be scary — they're only about the size of a turkey, so he modeled his dinosaur character on the Dinonychus instead.

Using the wrong name like that is called artistic license. Writers and Producers do it all the time, and it's why you can't always trust what you see in movies and on TV. But Mr. Chriton wasn't completely wrong. Both raptors and deinonychus are from the same family of theropods living in North America.

This one is full sized. When they were first discovered, these fossils changed everything we thought we knew about dinosaurs. We used to think all dinosaurs were cold blooded, but this one seemed to be covered with feathers. It helped to develop the idea that birds evolved from Dinosaurs and that most Dinosaurs were probably warm blooded too.

Reptiles

A lot of animals from the time of the Dinosaurs have survived right up to today. As you walk along this section of the path, look for reptiles that were living among the Dinosaurs.

Watch out for the big Snake and lizards behind the bench! These kinds of animals were around during the time of the Dinosaurs too. But like the insects and amphibians, they survived the big extinction event and are still here today.

Across the creek you'll see a big turtle. Turtles are one of the oldest reptile groups. They've been around much longer than dinosaurs and could get up to ten feet long and weigh over two tons. Their shells grow out of the spine along its back in a way that's similar to how a stegosaur's plates grow. Because of this, some people think turtles and stegosaurs could be related.

Crocodiles have also been around since long before the first dinosaurs. The largest crocodiles could grow to 40 feet long, about twice the size of a modern crocodile.

After the dinosaurs died off, crocodiles ruled the Earth for about ten million years until Mammals got big enough and vicious enough to take over. Crocodile fossils have been found all over the world including the high arctic, where polar bears live today.

Minmi Ankylosaur

The Ankylosaur is another plant eating dinosaur. They have been found on every continent except for Africa.

This is the minmi sub-species of ankylosaur. The regular ankylosaurs were about the size of a typical SUV, They all had armor plating for protection and spikes along the sides of their heads and necks. Like so many other plant eating dinosaurs, it had both a beak and teeth, and there is evidence that they had long flexible tongues as well. These animals were closely related to the stegosaur family of dinosaurs but they continued to thrive millions of years after the stegosaurs went extinct.

Pterodactyl and Velociraptor

Look at that little guy down there. Believe it or not, that's a velociraptor. It's not a baby, this is about how big they really were. Raptors had hollow bones so they were lightweight and fast. They could probably sprint to about 40 MPH, and were covered in feathers. It's not hard to imagine how a fast, lightweight feathered dinosaur might have become airborne. This is why we believe birds evolved from dinosaurs.

The velociraptors were smart too. They had the largest brain/body ratio of all the dinosaurs. But don't get the wrong idea. Compared to modern animals these were still pea-brained creatures. This one has come back to see what happened to his friends, but he hasn't noticed the pterodactyl perched in the branches over his head.

Pterodactyls are not true dinosaurs. They were a kind of flying carnivorous lizard swooping down to attack their meals from above. Most pterodactyl wingspans were only two or three feet wide, and the largest one ever found was eight feet. On the ground they could walk on all fours using the mid-point of their wings like forelimbs. In the air, that bulbous snout and large crest were probably used like a rudder. Wind tunnel tests on models have confirmed this idea. If you thought they should be larger, you're probably thinking of the pteranodons. Pteranodons were much larger and came a few million years after the pterodactyls.

T-rex

The Name Tyranosaurs Rex is from ancient latin and greek, meaning Tyrant Lizard King.

Tyranosaurs Lived in North America and were the largest carnivorous animals ever known to exist. They could grow up to 40 feet long and about 12 feet tall, about twice as big as this one. But not all of then were that big. That's just the largest specimen ever found. Most were smaller than that.

They only lived to be about 30 years old and may have had feathers, like a giant deadly chicken. Oh yeah, that dinosaur movie was wrong about these animals too. Judging from the muscle attachments to the bones, they could only run about 15 to 18 miles per hour-- still faster than most people.

T-rex was one of the last dinosaurs to die off in the great extinction event that killed all the dinosaurs. One of the big mysteries of T-rex is how they used their arms. They are unusually strong— they could lift over 400 pounds based on how the muscles attach to the bones. But they were too short to raise food up to the mouth. What do you think they were for?

Apatosaurus

This is a baby apatosaurus. But it's ok if you want to call it a brontosaurus or brachiosaurus, or diplodocus or even titanosaurus. Those species were all part of the sauropod family, and unless you're an expert, they all look pretty much the same— just like the different species of elephants or giraffes look the same to non-experts.

All the sauropods have similar features. Long neck, small head, long whip-like tails and big bodies. *Very* big bodies. The largest one ever found was 130 feet long and 60 feet tall. That's longer than three school busses parked end-to-end and as tall as a six story building. In comparison, the large sauropods were about four times as tall as a giraffe and more than nine times heavier than the largest elephant.

But not all the sauropod species were that big. The smallest adult sauropod is a species called the Magyarosaurus. It only grew up to 20 feet long and only six feet tall, about the size of an average pick-up truck.

We mentioned that this is a baby apatosaurus. The little one on the hill is only a few weeks old. The bigger one is only a year old. In ten years he'll be full sized, about ten times as big as this. These animals could live for about 100 years unless something else killed them first. But that's unlikely. There is evidence from skin impressions that they may have been covered with small armored scales, and their sheer size makes them difficult to attack, in the same way an adult elephant is almost impossible for lions to attack.

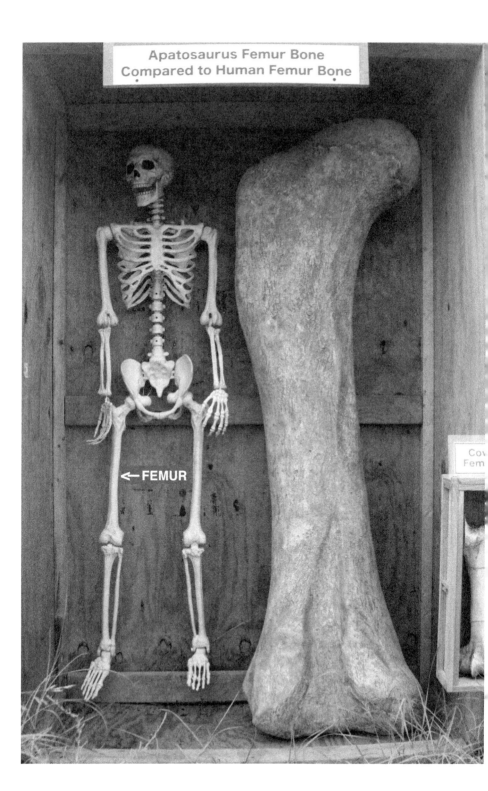

The reason for so much diversity is because the sauropods were such a successful family of dinosaurs. They have been found on every continent, including antarctica, and existed for about a hundred and fifty million years—from 210 million years ago until the end of the dinosaur reign in the big extinction event 65 million years ago. As they migrated from one climate zone to another, and as the planet changed over the millions of years they existed, these creatures also adapted and their bodies changed to match the conditions.

We can learn a lot about how an animal lived just by looking at it's bones and teeth and comparing them to what else we know. And the more we know, the more we can figure out.

Sauropod Eggs

Visit SpaceTime Park In New Braunfels, TX.

A family owned and operated outdoor science museum

Directions:
Address: 1001 Watson Lane East, New Braunfels, TX. 78130.
We are located halfway between San Antonio and Austin.
From Interstate 35, take exit 195 toward Watson Lane.
Follow the signs to Watson Lane EAST.
We are one mile from the freeway.

Be Prepared:
- Wear good walking shoes.
- Sunglasses, a hat and sunscreen are a good idea too.
- Feel free to bring your own food. Picnic areas are available.
- We'll also have water, cold drinks and snacks available.

What to expect:
You'll see more than twenty life-sized, highly detailed statues of animals that were alive at the time of the dinosaurs, spaced out on a quarter mile of trails. Each exhibit includes an automated narration to tell you about the animals and challenge some common misconceptions people have.

In addition to the Dinosaurs, you'll walk through an exact 1/250,000,000 scale model of the solar system, sized so you can walk at the speed of light. See an eclipse at the Earth pavilion! Every planet has is an information panel full of facts, photos and stories about what we know and how we know it.
And of course, we're adding more exhibits and more features every chance we get. It's a truly unique experience.

www.SpaceTimePark.com

CPSIA information can be obtained
at www.ICGtesting.com
Printed in the USA
LVOW06s0021300817
546903LV00034B/1662/P